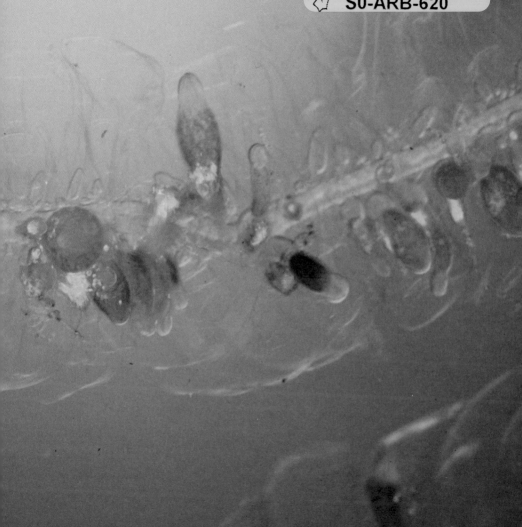

Amazing JELLIES
Jewels of the Sea

ELIZABETH TAYNTOR GOWELL

BUNKER HILL PUBLISHING
IN ASSOCIATION WITH
THE NEW ENGLAND AQUARIUM

First published in 2004 by Bunker Hill Publishing Inc.
285 River Road, Piermont, NH 03779 USA

10 9 8 7 6 5

Library of Congress Cataloguing in Publication Data
is available from the publisher's office

ISBN 978 1 59373 020 8

Designed by Louise Millar

Printed in China by Jade Productions

New England Aquarium

The New England Aquarium opened in 1969 to present, promote, and protect
the world of water. In addition to its programs and exhibits, enjoyed by more
than 1.3 million visitors every year, the Aquarium's conservation and research
programs are among its most important initiatives. This book reflects the
theme of a special exhibit of the same name that opened in April, 2004.

Endpaper: Siphonophores like this Praya *sp. are hydrozoans, found in colonies composed of
modified polyps and medusae, some for feeding, some for reproduction, and others for self-
defense. The Portuguese man-of-war is a siphonophore.*

*Ttitle page: This deep, reddish-brown jelly (*Periphylla periphylla*) is among the most
striking and widely dispersed deep-sea scyphomedusae in the oceans.*

Introduction

It floats in the water like a spaceship from another world, then sinks slowly down into the sea. Deep purple bands mark its umbrella-shaped body. Eight long and slender pink tentacles hang down from the edges of its body, and a frilly mass of pink and white, like ruffles on a dress, trails out from the center. As the animal sinks, it shimmers with rainbow colors in the sunlit water. A tiny fish larva swims by. It touches one of the tentacles and is instantly paralyzed—fallen prey to this strange and colorful creature of the ocean.

The bell of this beauty, called the purple-striped jelly (Chrysaora colorata)*, can reach a maximum of 2.3 feet in diameter. This species lives primarily off the coast of California, where it subsists on a diet of larval fish, copepods, fish eggs, and other jellies.*

Sea jellies like this one are among the most beautiful and unusual animals on earth. Jellies are found in all the oceans of the world and even in some freshwater lakes and rivers. In the summer, you might see some washed up on a beach or floating in a harbor. And if you could peer into the icy waters of the Arctic or Antarctic Oceans, you would see them there, too. Sea jellies have even been discovered more than 9,843 feet (3,000 m) below the surface, and they probably exist to full ocean depth.

*Even if you are not a swimmer, you can find jellies like this "by-the-wind sailor" (*Velella velella*) washed up on the shore. This particular kind of jelly spends its adult life floating at the surface, blown by the wind and carried by currents.*

*Not all jellies live in the ocean. This freshwater species (*Craspedacusta sowerbii*) is a hydrozoan thought to have originated in Amazonian waters; from there it has been introduced to other countries around the world.*

*This deep-sea jelly (*Periphylla periphylla*) can be found in all the world's oceans. It reaches depths of up to 23,000 feet (7,000 m) and catches fish and crustaceans with upward-floating tentacles that transfer captured food to the mouth below its bell.*

5

Many sea jellies belong to a group of animals with the scientific name **Cnidaria** (from the Greek *knide*, meaning nettle, a plant with stinging thorns). Sea anemones, corals, hydroids, the Portuguese man-of-war, and what used to be called "jellyfish" are all animals in this group. Scientists estimate that there are more than 9,000 species of Cnidaria. All of them share common characteristics: jellylike bodies, tentacles, and stinging cells. But these animals are also so different from each other that scientists have split Cnidaria into three separate subgroups: the **scyphozoans, anthozoans,** and **hydrozoans**. Most scientists now agree there should also be a fourth subgroup, the box-shaped **cubozoans**.

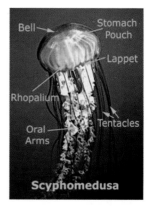

The umbrella-shaped body of a scyphomedusa, called the bell, contains a stomach pouch, lobes along the margin of the bell, called lappets, and the rhopalium sensory structure that helps the jelly detect light, telling it which way is up and which way is down. The oral arms trap and transfer food to the jelly's mouth and stomach.

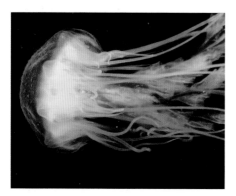

Lion's mane (Cyanea capillata), a scyphozoan

Moon jelly (Aurelia aurita), a scyphozoan

Coral polyps, an anthozoan

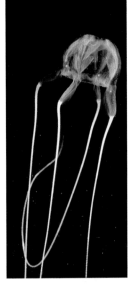

Box jelly (Carybdea marsupialis), a cubozoan

Coral polyps, an anthozoan

Rose anemone (Urticina piscivora). an anthozoan

The scyphozoans are the most familiar jelly animals—free-swimming, bell-shaped animals best known as "jellyfish." Scientists prefer to call these jelly animals **medusae** (**medusa** in the singular), because they are not fish at all, and that is how we will refer to them in this book. In Greek myths, Medusa was a monster whose head was covered with writhing snakes instead of hair. The many tentacles surrounding the body of the jelly may have reminded early scientists of this mythical creature.

Anthozoans look like flowers or upside-down bells on stalks. The body of an anthozoan jelly is called a **polyp**. Examples of these animals are corals or sea anemones. While the medusa jellies are free-swimming, and the anthozoan jellies are attached polyps, the jellies of the third cnidarian subgroup—the hydrozoans—can be either free-swimming or polyps. **Hydroids** and freshwater **hydra** are examples of polyp hydrozoans. Both hydroids and hydra form colonies in the shape of graceful feathers or tiny branching trees. The beautiful but dangerous Portuguese man-of-war is a special floating form of hydrozoan.

Caravaggio's painting depicts the Medusa in Greek myth as a monster whose snake-infested head may have reminded scientists of the tentacles of the sea jellies from the group called schyphozoans.

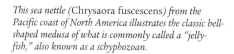

This sea nettle (Chrysaora fuscescens) from the Pacific coast of North America illustrates the classic bell-shaped medusa of what is commonly called a "jellyfish," also known as a schyphozoan.

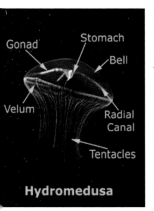

Hydromedusa

Labels: Gonad, Stomach, Bell, Velum, Radial Canal, Tentacles

The free-swimming medusa stage of a hydrozoan is called a hydromedusa. Beside familiar features such as the mouth, bell, and tentacles, are the organs that contain male or female reproductive cells (gonads); the muscular structure called the velum that helps the jelly move; the cavity (stomach) called the manubrium that ends in the jelly's mouth; and a water circulatory system called the radial canal.

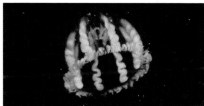

The anemone shown here with an anemone fish is an example of an anthozoan. Its body consists of a "polyp" attached to a surface on the ocean bottom.

Although these are not the same species of hydrozoan, they represent the bottom-dwelling attached stage (top two photos) and the free-swimming medusa stage of hydrozoans. Some hydrozoans live entirely as free-swimming medusae, others live only as attached polyps, and there are many that go through both stages. Hydrozoan medusae, however, are usually much smaller; some are only half the size of a thumbtack.

9

Some of the ocean's strangest jelly animals are not cnidarians at all. Ctenophores, or comb jellies, exist in a variety of shapes. Some are called sea gooseberries, others sea walnuts because of their appearance. The name "comb jelly" comes from the animal's eight comb plates, which contain fused hair-like structures called cilia; each individual hair or cilium is just one tooth of the comb. Found on all comb jellies, the cilia constantly move back and forth to propel the jellies through the water. The close spacing of the comb teeth can break light into rainbow colors. Unlike cnidarians, comb jellies do not sting at all.

Salps are oceangoing members of the scientific group of animals called **tunicates**. Unlike other bottom-dwelling tunicates called sea squirts, salps do not produce larvae. They also lack the primitive spinal chord that would make them cousins of vertebrates.

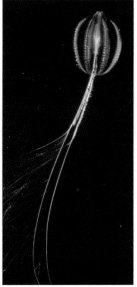

*Sea gooseberry photos
Members of the comb jellies, sea gooseberries, trail tentacles behind them as they move through the water. The tentacles capture prey because they are sticky, not because they bear poisonous nematocysts.
The sea gooseberry shown in this picture has trapped food on one of its tentacles. It is named for its resemblance to the fruit that birds eat.
Unlike other bottom-dwelling tunicates called sea squirts, salps lack the primitive spinal chord that would make them cousins of vertebrates. Salps do not produce larvae.*

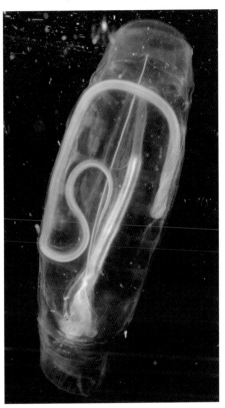

Comb jellies are named for the comb plates that run in strips down their bodies like the teeth of a comb. These teeth can refract (break) light into rainbow colors.

Ctenophores, or "comb jellies," snare their prey on sticky cells, while salps like the one shown here (Helicosalpa virgula) trap plankton in a mucus net as water passes through their bodies. The paperclip shape inside this salp is a chain of new salps being formed.

Salps use bands of circular muscles to force water through their bodies. Salps do not sting.

11

Salps eat plankton, tiny floating ocean plants and animals. They suck water into their mouths by contracting circular muscles located all around their barrel-shaped bodies. A mucus net inside the salp's body filters tiny organisms from the water. The water is then released through an opening at the other end of the animal. This squirting action jet-propels the salp forward.

Like some other jellies, such as hydroids, salps are found in both individual and colonial forms. Some beautiful colonial salps are mostly transparent with a few brightly colored marks that look like stars. These salp colonies are made up of one parent salp along with many new salps that have budded off from the first. Budding salps often form chains that can be 6 feet (1.8 m) long or longer, made up of dozens of individuals. In motion, a giant salp chain twists and twirls like a great ocean snake.

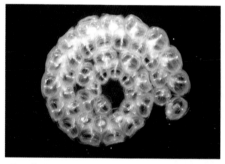

This salp (Pegea socia) *is an efficient filter feeder. A colony of these animals may contain as many as 150 to 200 individuals.*

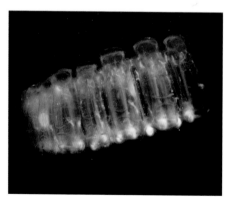

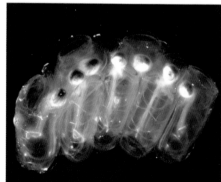

A closer view of Pegea sp. *shows the individual salps that are part of a colony.*

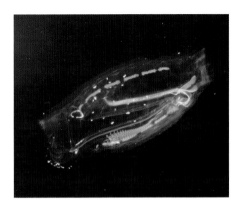

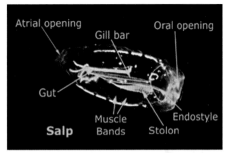

Atrial opening

Gill bar

Oral opening

Gut

Salp

Muscle
Bands

Stolon

Endostyle

*Top left: The solitary salp (*Cyclosalpa bakeri*) lives in both tropical and temperate oceans where, at night, it rises to the surface to feed.*

Top right: The endostyle produces mucus that traps food from the water circulated through the salp's oral opening. The stolon is a chain of new salps produced asexually. The atrial opening is where water leaves the salp after it has entered at the oral opening.

*Right: The mucous-feeding net of this salp (*Pegea *sp.) has been dyed red from trapped carmine particles.*

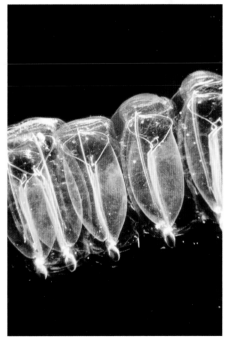

13

Jelly Forms

Sea jellies come in all sizes. Imagine a jelly as wide as a backyard satellite dish. *Cyanea*, or the huge lion's mane jelly, is a real sea giant, as much as 12 feet across (3.6 m) with tentacles 100 feet (30 m) long. Other jellies, such as the hydroid *Obelia*, are full grown at 1/10 of an inch (3 mm). The moon jelly *Aurelia* is usually 6 to 10 inches (10 to 25 cm) across, about the size of a dinner plate.

Medusa jellies were once called jellyfish, but they are not really fish at all. Fish are **vertebrates**, meaning they have backbones, and fish have gills and scales. The medusa and the rest of the sea jellies are **invertebrates**—they have no backbones. They also lack gills and scales.

Sea jellies are very simple animals. They are 95 percent water, with a little bit of protein, oil, and salt. Most have a hollow body shaped like a bell or umbrella. The body is made of two thin layers of cells with a thick layer of jellylike material in between. The mouth is an opening at the center of the animal. Medusa jellies generally have tentacles around the edges of the bell or the umbrella. Some of the medusa jellies also have frilly arms around their mouths.

Jellies have lived on the earth for millions of years. Jelly fossils have been found in rocks

The bell of the egg yolk jelly (Phacellophora camtschatica) *above can measure up to two feet across.*

500 million years old. These exposed rocks are located in Germany, the United States, and Canada. The oldest well-known jellies are ctenophores, also known as "comb jellies," which are ~520 million years old; but the oldest-known true or medusa jellies are from the Burgess Shale, which is ~515 million years old. Paleontologists (scientists who study fossils) believe that the fossils were formed when dead jellies sank to the soft mud bottom of an ancient sea and were covered with fine layers of mud and silt. The fossils we see today are impressions left by the jelles' bodies.

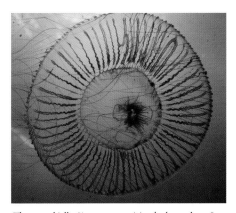

*The crystal jelly (*Aequorea sp.*) is a hydromedusa. It contains proteins, including one that is green and fluorescent that used to attract commercial harvesting for use as a marker in biomedical research.*

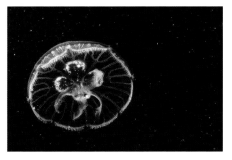

*The moon jelly (*Aurelia aurita*) makes it easy to realize that sea jellies are 95 percent water.*

*Compared to the egg yolk jelly on the previous page, the tiny polyps that line the tree-like branches of this hydroid (*Obelia*) are mere fractions of an inch.*

This sandstone deposition from the Wonewok formation in the state of Wisconsin in the United States is about 500 million years old. It shows the undersides of large and small medusae stranded in an intertidal setting.

15

How Jellies Work

Sea jellies are among the simplest animals in the world. They do not have many of the organs that most complex animals (including humans) have. They have no brain, no liver, no kidneys, no heart, no lungs, or gills. Sea jellies do not need blood because oxygen and nutrients don't have to be transported long distances to their body cells. They absorb oxygen directly from the water as it passes through the thin walls of their body tissues.

The mouth of the medusa jelly is located on the underside of its body, in the center. It is the only opening into and out of the medusa's stomach, a hollow cavity where food is digested. Like oxygen, nutrients pass directly into the cells surrounding the stomach and through the jelly layer to other parts of the body. The frilly arms some medusae have around their mouths help transfer food to the stomach. Once the food is digested, wastes are passed back out through the mouth.

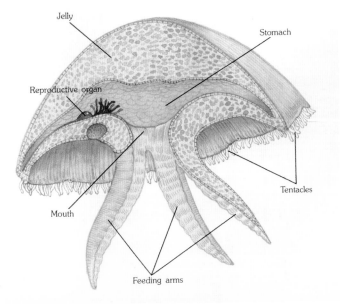

A view of the underside of a moon jelly shows many of its survival secrets. There are sensory structures along the bell margin that contain light- and gravity-sensitive cells. These help the jelly to detect sunlight and sense whether it is upside down or turned sideways. A dense ring of tentacles and frilly oral arms enable the jelly to catch and feed on tiny animals.

Jelly

Stomach

Reproductive organ

Tentacles

Mouth

Feeding arms

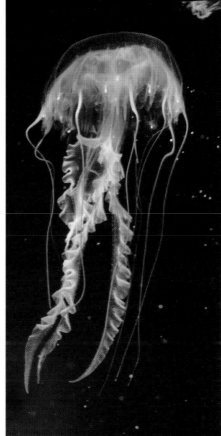

*The hydromedusa crystal jelly (*Aequorea sp.*) above stands in contrast to the long- and frilly-armed scyphomedusa as represented by the Pacific sea nettle below (*Chrysaora fuscescens*) and the Atlantic sea nettle (*Chrysaora quinquecirrha*) to the right.*

17

Getting Around

A medusa swims by contracting muscles around the edge of its body. As the muscles contract, water is pushed out of the hollow body, and the medusa is jet-propelled in the opposite direction. Squids and octopuses use the same technique for making fast get-aways. Although this may not seem like much of a swimming technique to us, a medusa can travel long distances just by pulsing its body. One species of Mediterranean medusa, *Solmissus albescens*, is only about 1¹/₂ inches (3.75 cm) across. It swims up and down through the **water column**, a distance of 3,600 feet (1,100 m) a day, following the tiny sea animals, or **zooplankton**, on which it feeds. This distance would be equal to a 33-mile (53-km) swim for a six-foot (1.8m) person.

Although medusae can move very well up and down in the water column, their swimming technique is no match for strong currents, wind, and waves. At certain times of year, beaches are littered with the bodies of medusae washed up on shore by storms or strong seasonal winds. Once on shore, the medusae have no way to get back to the sea. Their jellylike bodies dry up, and only filmy circles are left behind on the sand. Some close relatives of the medusa, such as the Portuguese man-of-war, have floats to keep them at the ocean's surface. These animals don't really swim, but drift wherever they are taken by the wind.

*Comparing the Atlantic sea nettle (*Chrysaora quinquecirrha*) with the squid shown at left illustrates many similarities between the two. Contracting muscles around the edge of the jelly's body forces water out of its bell and pushes it forward. When they need to travel quickly, squid move in much the same way, by forcing water out of their main body cavities.*

*Some kinds of jellies travel great distances and are surprisingly strong swimmers. Ocean winds, currents, and tides, however, are sometimes more powerful than the jellies, and the animals are washed up on shore. Both the by-the-wind sailor (*Velella velella*) shown in a small cluster above, and in a mass on the beach below, and the Portuguese man-of-war drift on the ocean surface. They are even more vulnerable to beach strandings.*

How Do Jellies Sense Their World?

Do sea jellies feel, hear, smell, taste, and see their undersea world? Unlike more complex animals, their senses are limited. Sea jellies have touch receptors on their tentacles and around their mouths to help capture food. These touch receptors may also detect vibrations in the water caused by the movement of a fish, crab, or other animals swimming by.

Sea jellies do not have a nose or tongue. They have special cells that smell and taste scattered all over their bodies. Sea jellies do not have eyes like human eyes, but many have light-sensitive organs around the margins of their bodies. In most cases, these organs do not detect shapes or movement, but allow the jelly to tell light from dark. Jellies can tell up from down by sensing the sunlight at the surface of the ocean.

Sea jellies also stay upright with the help of special balancing organs, which are located along the outside edges of their bodies. These fluid-filled sacs each contain a tiny, stonelike granule. When the sea jelly tilts too much to one side, the granule touches cilia, which stimulate nerve endings. These nerves cause the jelly's muscles to contract, putting it back on course. In a similar fashion, we have tiny stones in our inner ears that move around as our heads tilt, sending signals to the brain that help us keep our balance.

*The Palauan lagoon jelly (*Mastigias *sp.) hosts photosynthetic algae. These jellies don't just wait for the sun—they chase it. In lakes in Palau, they migrate from west to east, carefully avoiding shadows.*

*Like other scyphozoan jellies, the Pacific moon jelly (*Aurelia labiata*) contains pouches, called statocysts, along the edge of its bell. The pouches are lined with sensory hairs that provide a sense of balance and orientation for the jelly.*

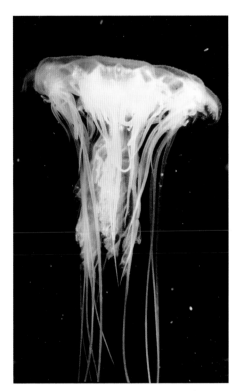

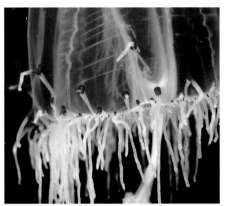

*Above: The rim of the bell of the lion's mane jelly (*Cyanea capillata*) contains sense organs for balance and orientation in the water column.*

*Right: Light-sensitive spots, each with a ring of red pigment, are situated at the bases of the smaller tentacles of this hydromedusa (*Scrippsia pacifica*). Also called bell jellies, they will react to night lights, such as that from a camera flash, and go in the other direction.*

21

The Jelly Life Cycle

Just as a frog must spend time as a tadpole before it grows up, sea jellies go through several life stages before they reach adulthood.

Let's look at the moon jelly *Aurelia* as an example. An adult moon jelly has a classic, bell-shaped medusa form. Some moon jellies are male. Some are female. The females produce eggs and hold them on the frilly arms around their mouths. The male jellies produce sperm and release them into the water. The females then take in the released sperm to fertilize their eggs. Each fertilized egg stays on the female jelly's mouth arms, where it grows into a tiny, pancake-shaped **larva**. The female moon jellies then release their larvae to the ocean, where they ride the currents and waves until they reach a rock or other hard surface on which they settle. Unlike their floating, free-swimming parents, the baby jellies need to attach themselves to a hard surface under water, and stay there.

Once settled, each larva transforms itself into a polyp. The polyp, which at first looks like a tiny sea anemone, uses tiny tentacles to capture zooplankton for food. As it grows, the polyp may produce buds, or branches, that eventually break off and become separate polyps. These budded polyps are entirely new animals.

The images shown here from top to bottom illustrate the various stages of growth and development of the moon jelly (Aurelia aurita). They begin with the attached stage (photo upper left), which reveals a miniature jelly about to bud off. Then comes an immature medusa stage, called an ephyra, shown here in the photo above right. The bottom photo is a young moon jelly.

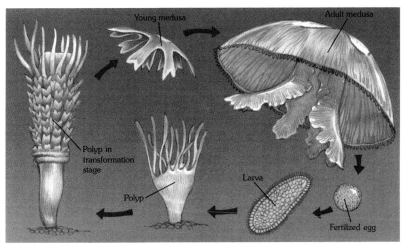

The life cycle of a scyphozoan

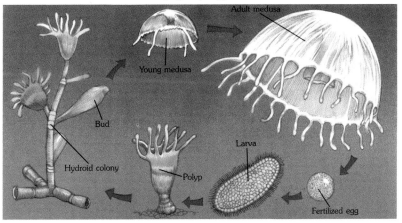

The life cycle of a hydrozoan

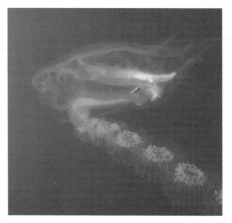

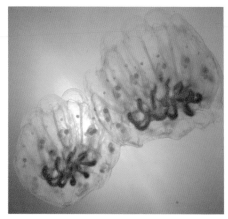

A solitary salp (Cyclosalpa affinis) *produces new salps asexually by forming linked whorls that trail off and are released into the water. Top right: Closeup of the linked whorls.*

A Christmas tree hydroid shows the classic form of the attached bottom-dwelling hydrozoan with its branches containing tiny polyps.

A hydromedusa stage of another kind of hydrozoan not related to the Christmas tree hydroids. Not all hydrozoans have both swimming and stationary phases.

The attached stage of the egg yolk jelly shows the budding disks that will, one by one, float off freely into the water column.

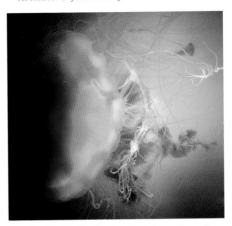

Once the egg yolk jelly has gone through its polyp and larval stages, it matures into the adult shown here.

After several months, or sometimes several years, the polyps start to undergo an amazing transformation. Grooves form in the body of each polyp. The grooves become deeper and deeper until they go all the way through the polyp's body, creating a stack of tiny disks. Each of these disks is a baby moon jelly. These tiny jellies, each only 1/8 inch across (0.3 cm), gradually break off from the stack, one by one, and swim away on their own. Those that survive become the next generation of adult moon jellies.

From each larva that survives to reproduce, many moon jellies eventually develop—some from new polyps that bud off from the original polyp, and others from the original polyp itself.

Unlike the medusa jellies, comb jellies are simultaneous hermaphrodites—each contains fully functional male and female parts. They generate a free-swimming larvae, which for some varieties is an exact miniature of its parents.

Salps also reproduce without "settling down." Solitary salps produce chains of new salps through budding. These chains then reproduce sexually, often with each salp breeding another solitary salp.

Stinging

Jellies are beautiful to look at, but most people don't want to get too close—cnidarian jellies sting!

Some stings are so mild you might not feel them. Some, however, are painful and can cause sharp, burning sensations and red, swollen welts.

The stinging power of cnidarian jellies comes from special cells on their tentacles and other body surfaces that contain capsules called **nematocysts**. Many nematocysts have trap doors with tiny hairlike triggers. Coiled inside the capsule is a long, hollow tube. When the jellies' prey, or a swimmer's leg, touches the trigger, the trap door opens and the tube is turned inside out as it extends from its capsule. The whole event takes place in milliseconds.

The nematocyst tubes of some stinging jellies work like the hypodermic needles doctors use to give injections. When the tube pierces the victim's skin, a paralyzing toxin is injected. Other jellies have tubes with dozens of tiny hooks that stick onto their prey like Velcro. Some even work like tiny lassos that wrap around and trap tiny animals. Scientists have found 30 different types of nematocyst tubes—some with barbs, some smooth, some short, and some

This man-of-war has pulled its nematocyst-covered tentacles back toward its body. When undisturbed, these same tentacles hang down to catch unwary prey.

A close-up view of the tentacles of an egg yolk jelly (Phacellophora camtschatica).

300 times as long as the nematocyst capsule is wide.

Not all jellies sting. Comb jellies snare their prey on sticky cells, while salps catch plankton in a net of mucus as water passes through their bodies. Neither has nematocysts, and neither can sting at all.

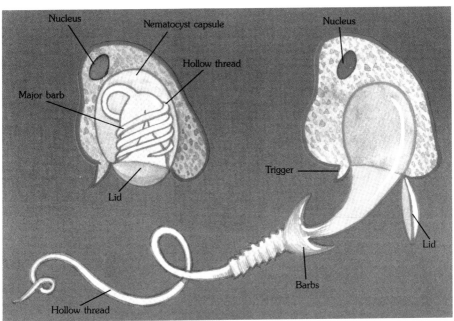

Nucleus

Nematocyst capsule

Hollow thread

Major barb

Lid

Nucleus

Trigger

Lid

Hollow thread

Barbs

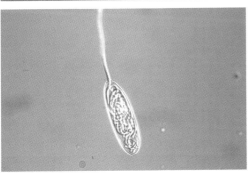

Above: Jellies from the family Cnidaria are armed. They have specialized stinging cells called nematocysts. Triggered by touch or a chemical cue, a long hollow tube is thrust out from each nematocyst capsule. If the tube pierces a swimmer's skin, a painful toxin is injected. The above drawing shows the stinging cell of a sea jelly before and after firing the nematocyst.

Left: A magnified view of a nematocyst from an anemone shows the barb still coiled up and ready to fire.

27

Prevention and Treatment

How do you avoid being stung by a jelly? First, get to know the sea life in your area. Stay away from beaches when large numbers of stinging jellies are blown ashore by high winds. After a storm, look around to see if stinging jellies are in the water. Tentacles torn off an injured jelly can still sting. A dead jelly washed up on a beach should be avoided. The stinging power of nematocysts can remain even after the jelly is dead.

In Australia, lifeguards used to wear pantyhose to the beach—one pair on their legs and one pair with holes cut in it to fit over their heads and hands, on their upper body. They might have looked funny, but the fabric protected them from the short but deadly nematocysts of the sea wasp.

Today, Lycra bodysuits have been designed for the same purpose. Some swimmers also coat their bodies with Vaseline or lanolin to prevent jelly stings.

What do you do if you are stung? First, get out of the water. Researchers believe, that in a few cases, good swimmers have drowned because of jelly stings. Once on land, try to remove the tentacles, being careful not to get stung again. Do not try to rinse the tentacles off by washing the skin with water. This could cause more nematocysts to fire. Do not brush or rub the tentacles off. This will also cause nematocysts to fire. Some people have severe allergic reactions to even mild jelly stings, so it is best to see a doctor as soon as possible.

*Some kinds of jellies are especially memorable for their sting. On the Pacific coast of North America, the sea nettle (*Chrysaora fuscescens*) is often the culprit.*

If you were swimming off the north Atlantic coast of the United States and were stung, the stinger would probably be *Cyanea*—the lion's mane jelly; in the Chesapeake Bay area, it would be *Chrysaora*—the sea nettle. South of Virginia Beach, along the Gulf of Mexico, on the west coast of the United States, or in Hawaii, the stinger would likely be *Physalia*—the Portuguese man-of-war. *Physalia* also occur in the Caribbean and Mediterranean seas.

The most deadly jelly animal is the sea wasp. This creature lives in tropical waters throughout the world. In the southeast Pacific and Indian Oceans, one species, *Chironex fleckeri*, is believed to have caused about sixty deaths over the past hundred years.

In American waters, the most dangerous species of sea wasp is *Chiropsalmus*, which is seen in the Gulf of Mexico and off the southeastern shore of the United States. A sting from this jelly can make it hard to breathe, and the victim might need to be taken to a hospital for treatment. But *Chiropsalmus* is not as dangerous as the Australian sea wasp. The purple jelly (*Pelagia noctiluca*) has been a problem when it occurs in large numbers in the Mediterranean Seas.

Although some species of sea jelly can be very dangerous, only a few (perhaps up to 70 or more species) of the over one thousand described have nematocysts numerous or powerful enough to give humans a noticeable and significant sting.

*The sea wasp (*Chironex fleckeri*) has tentacles up to 9 feet (3 m) long, and they can be deadly. A close encounter with this jelly can cause death to a human swimmer in a matter of minutes.*

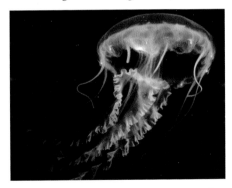

*The jelly with one of the most bothersome stings on the Atlantic coast of North America is the Atlantic sea nettle (*Chrysaora quinquecirrha*).*

29

What's for Dinner?

Most jellies eat tiny floating animals called zooplankton, which includes the eggs and larvae of fish. They also compete with juvenile fish eating the same food. Some jellies eat adult fish and even other jellies. Most cnidarian jellies fish for their food with tentacles that may be many times longer than their bodies. The Portuguese man-of-war has tentacles up to 60 feet (18 m) long. When an unlucky fish or tiny crab touches one of these, the nematocysts shoot out and the victim is trapped. The man-of-war then contracts its tentacles, reeling in the captured prey. After a nematocyst is fired, the jelly's body produces a new one to replace it.

The insect-like copepods swimming in this jar are an example of the foods jellies prefer. The jar contains thousands of the tiny animals.

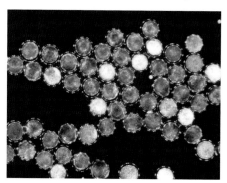

Animal plankton such as these eggs of shrimplike invertebrates, called euphausids, are among the favorite foods of sea jellies.

This larva of a fish called a grunt sculpin, shown magnified here, is another example of the kinds of foods sea jellies consume as they float in the ocean currents.

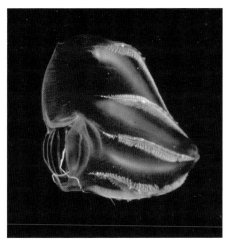

*Tentacles of the man-of-war (*Physalia *sp.) can reach 60 feet in length in some of the longer-living specimens.*

A comb jelly seizes another smaller jelly, called the sea gooseberry, for its next meal.

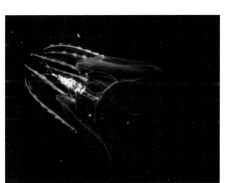

*The gut of this comb jelly (*Bolinopsis infundibulum*) contains a feast of tiny invertebrate animals call copepods.*

One kind of comb jelly has just consumed another, a smaller species visible inside its gut.

31

Who Eats Sea Jellies?

There are a number of ocean animals that find jellies, even the big stingers, tasty prey. The ocean sunfish sucks down jellies with a noisy slurp. And it must eat a lot of them, since these fish can grow to a whopping 1,500 pounds (680 kg) eating almost nothing but jellies.

Sea turtles also find jellies a tasty treat. Loggerhead sea turtles are especially fond of the Portuguese man-of-war. When winds or currents bring huge numbers of these jellies together in one part of the ocean, sea turtles gather and feast on the purple floats. Between bites, they brush away the stinging tentacles with their flippers, but their sensitive eyes still swell from the stings.

One jelly predator is a small blue sea slug, or nudibranch, called *Glaucus atlanticus*. This animal lives off the coast of Australia and feeds on *Physalia* and other sea jelly tentacles. When swimmers reported that they had been stung by this little nudibranch, scientists investigated and discovered unexploded man-of-war nematocysts in the nudibranch's tissues. It appears that the nudibranch can move stinging capsules from the man-of-war into its own body, thus equipping itself with stolen weapons.

Some jellies even eat other jellies. One jelly predator can grow to be as much as a foot (30.5 cm) long. It is the comb jelly *Beroe*. Like other comb jellies, *Beroe* does not sting. It does not even have tentacles. Basically, this jelly is a swimming mouth that opens wide to vacuum in its prey, usually smaller species of comb jellies. Inside its mouth, *Beroe* has specialized structures that are shaped like fangs or the teeth of a saw. These are used to trap or take bites out of the prey.

Scientists might be underestimating how many animals eat jellies. Since jellies are digested so quickly and completely, they don't appear in the stomach contents of other animals. But many species eat jellies, even as a secondary food source. Among jelly predators: sea turtles, tuna, ocean sunfish, butterfish, spiny dogfish, and other jellies.

In some parts of the world, jellies are enjoyed as a delicacy. In Japan, China, and Korea, people eat salted and dried jellies that look like thin white pancakes and are chewy, like rubber bands. Other types of jellies are soaked in salt water, steamed, then seasoned with spices and served.

Despite the stinging power of this sea nettle floating just below the surface, the northern fulmar seems to disregard the nematocysts and feasts without consequence.

Rockfish nibble on the tentacles of the egg yolk jelly with apparent impunity.

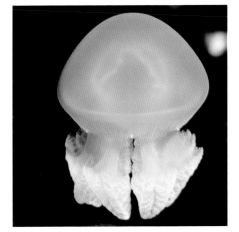

Blue jellies may be a nuisance to fishermen, but they and other species of jellies are considered a dining delicacy in some parts of Asia.

Given the opportunity, even anemones dine on jellies, like the one shown here enveloping an egg yolk jelly and pulling it into its mouth.

33

Deepsea Jellies

Among the most beautiful jelly animals are those people usually see only in photographs. They live in the cold, dark, deep sea, hundreds, even thousands, of feet beneath the surface. Scientists once thought that the deep sea was like a watery desert with little animal life. But that idea is changing. With new ways to explore the ocean's depths, we are finding that it is much richer in life of all shapes and sizes than we ever expected.

The deep sea is still a new frontier. Before the invention of SCUBA (Self Contained Underwater Breathing Apparatus) in the 1940s, scientists studying marine life could not venture underwater for any length of time. SCUBA enabled divers to carry tanks of air with them; observers could then stay underwater for up to an hour or more. But the depth of these explorations was limited to 100 or 200 feet (30 or 61 m). Today, small submarines, specially designed for deep-sea research, take scientists down to depths of 20,000 feet (6,100 km), although depths of 2,000 to 3,000 feet (610 to 915 m) are more common. Here researchers can explore an undersea wilderness about which very little is known.

One surprise for these deep-sea explorers was the number of jelly animals that they found—medusae, anemones, and corals of fantastic colors and sizes. In fact, scientists now believe that jelly animals may be one of the most common type of animal life in the ocean depths.

Scientists have also found that the deep sea is not always dark. Some deep-sea jellies can turn on bright lights. Like underwater lightning bugs, they give off sparks of blue, green, or blue-green. This living light, called **bioluminescence**, is caused by a chemical reaction inside the jellies. Scientists are not sure why jellies glow in the dark. It may be to startle or confuse predators, or to attract their tiny zooplankton prey. Many sea jellies that live on the surface also produce bioluminescence. Sea captains sometimes see the ocean surface glow at night with the light of hundreds of jellies disturbed by a passing ship.

Scientists use various techniques and vehicles to explore sea jellies of the midwater and the deep. The deep-sea submersible Alvin, from the Woods Hole Oceanographic Institution, is shown to the left. On the right, divers use a network of lines to keep together in the open ocean.

This photo shows the mouth end of Atolla vanhoef-feni, *thought to be widely distributed in the world's oceans. It is a deep-ocean jelly, living at depths of 1,600 to over 3,000 feet (500 to 1000 m).*

35

*Sometimes collected in midwater fishing trawls, this hydromedusa (*Colobonema sericeum*) is found in the Pacific, Atlantic, and Indian oceans. Living at depths of 1,600 to 4,900 feet (500 to 1,500 meters), if threatened it will cast off its bioluminescent tentacles.*

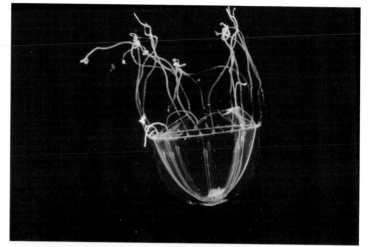

*Preferring deep water near the ocean bottom, this brilliantly colored hydromedusa (*Benthocodon pedunculata*) unfurls several thousand tentacles, shown here after they have been pulled back toward the jelly's body.*

Most deep-sea jellies are brilliantly colored. Purple and red are the most common shades. The reason for these brilliant colors is a mystery. For jellies that live at the surface, scientists know that pale or transparent colors act as camouflage, making jellies almost invisible to predators. But in the perpetual darkness of the deep sea, what role can bright colors play?

Perhaps the color is simply a beautiful by-product of what these jellies eat. Or it may be an adaptation to hide what the jellies have themselves eaten. Many of the tiny crabs, shrimp, and fish that deep-sea jellies eat also glow in the dark. If the jelly predators were transparent, their most recent meal would glow inside them like a giant underwater light bulb, making them easy targets for a hungry fish or squid. Instead, their dark body colors hide the bright lights of the animals they eat. A recently discovered deep-sea species of comb jelly, *Lampocteis cruentiventer*, is called "bloody belly jelly," because of its dark red stomach.

Although we are learning more about the deep sea, there are still limits to studying life in the ocean depths. First, scientists can go only as deep and as far as the research submarines that carry them. Second, on very deep dives, they cannot leave the vessels. Therefore, much of what we know about

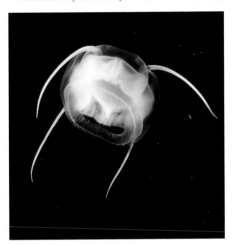

Found in most warm and temperate oceans, Aegina citrea *is an active swimmer that uses its four tentacles to capture other jellies such as ctenophores, salps, and hydromedusae. Its bell usually measures less than 4/5 of an inch (20 mm).*

deep-sea life is based on photographs and accounts of what scientists have seen from the windows of submersibles. Finally, although some research subs are equipped with mechanical arms and baskets for collecting scientific specimens, it is very difficult to maneuver this equipment to catch a moving animal. And jellies are very fragile—when touched, they break apart or are injured, making it difficult to bring them back alive and in one piece.

37

The Future of Jellies

For many years, jelly researchers were more interested in finding ways to get rid of these fascinating animals than they were in understanding them. Today, scientists appreciate the amazing things these simple creatures can do and are beginning to recognize the important role they play in the chain of life in the oceans.

What kinds of research do jelly scientists do? Some are continuing to search for an effective antivenin that will save victims of sea wasp stings. Others are studying the chemicals in medusae and other jellies for possible use in treating cancer and other diseases. One of the bioluminescent chemicals found in a medusa from the Pacific Northwest has already been found to be useful in certain types of medical research. This substance allows doctors to trace the movement of specific chemicals through the body.

Some researchers are discovering and collecting jelly animals that are new to science. Others are researching how these animals live—what they eat, how they reproduce, and what special features enable them to survive. This may help explain why jellies appear in swarms at certain times of the year and offer clues about how to control them.

One of the early challenges in jelly research was to develop laboratory equipment and techniques for keeping these unusual animals alive. Since it was too difficult to collect enough of the tiny animals jellies eat, scientists learned to grow food for the jellies. They also designed special round tanks, with no flat walls for fragile jelly animals to bump into. Now that these techniques have been perfected, many more types of jellies can be collected and studied. As this research continues, scientists are sure to discover many more amazing things about these animals and the ocean world in which they live.

Scientists are developing better techniques to collect jelly animals, in the hopes of understanding how these rainbows of the sea are adapted to life in the ocean.

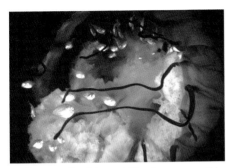

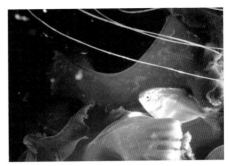

As the science of studying sea jellies advances, we will learn more about how jellies survive and how other animals live alongside them in the sea. Some jellies are like floating apartment houses. Small fish or invertebrates hide among their tentacles, safe from other predators. Scientists once believed these animals were immune to the jelly's sting. But they might also be very good at avoiding contact with the stinging tentacles.

A special round tank, called a kreisel, allows jellies to avoid both contact with sharp surfaces and getting sucked into the grid of circulation pumps.

39

Jellies and the Environment

Although jellies have been swimming through the world's oceans for hundreds of millions of years, today many jellies are greatly increasing in number. Some of these increases are likely due to natural population variations. Others are almost certainly caused by changing ocean conditions—changes caused largely by human activities.

Over-fertilization and Pollution

Fertilizers and manure can help our lawns stay green and our tomatoes grow big and red. When washed into rivers and oceans, however, the excess nutrients from fertilizers cause phytoplankton—tiny ocean plants—to grow and reproduce very quickly. This process is known as eutrophication, and can mean trouble for fish, shrimp, and many other creatures. When the millions of little phytoplankton die they use up oxygen in the water. Big fish, turtles, shrimp, and other fast-moving animals leave the area. But many creatures cannot escape, including baby fish, crabs, mollusks, and others. Jellies don't mind the low oxygen levels and feast on the smorgasbord of tiny oxygen-starved animals—zooplankton—that were not able to escape.

Litter often ends up in our rivers and oceans, where it can float around for years. Plastic bags and balloons floating in the ocean look just like jellies. While the plastic does not hurt the jellies, it causes problems for the animals that eat them. If a sea turtle mistakes a plastic bag for a tasty jelly snack, it could be in big trouble—plastic can actually kill the animals that eat it.

Over-fishing

People love to eat fish, and many of the fish that we eat dine on jellies. These are jelly predators. Other fish eat the same food that jellies eat—zooplankton. These fish are jelly competitors. When we remove fish from the oceans for our dinners, we are also removing the predators and competitors that normally keep jelly populations under control. Without predators or competitors, jelly populations can grow very rapidly.

A fast-growing jelly population is called a jelly bloom. These jelly blooms can make it hard for fish to recover from fishing pressure. Not only do jellies compete with young fish for the same food (zooplankton), but many jellies will also eat the young fish and fish eggs. The combination of starvation and predation pressures makes it very hard to be a fish during a jelly bloom.

Fertilizer runoff from watering fields like this one in Florida can get into the water table and run off into the ocean if there is shoreline nearby. Fertilizer, manure, and sewage in the ocean all decrease oxygen levels, leaving water hypoxic. Fish find it hard to breathe, so they go elsewhere. Fish larvae are slowed and can't escape predators. But jellies live comfortably. In fact, they live well—at low oxygen levels, jelly predators feast like kings on their stunned and oxygen-deprived larval prey.

Too many huge explosions of sea jellies, such as the Pacific sea nettles shown here, can compete with fishes for the resources of the ocean.

Global Climate Change

Global climate change is making our planet warmer and our summers hotter and longer. But this does not mean more days at the beach—the jellies may keep you out of the water. Many jellies are seasonal, which means that they will come and go with the warm-weather months. And longer summers mean more jellies.

Every year, jellies begin feeding once the water is warm enough to support them. As our oceans heat up, the jellies have been appearing earlier and earlier. And since the warm weather is lasting longer, the jellies stay longer than ever before. This is a serious problem for many fish. Jellies eat an incredible amount of food every day, and most of that food is zooplankton, including fish eggs and baby fish. If the jelly season is getting longer every year, then the jellies will be able to eat ever-increasing numbers of fish eggs and baby fish.

41

You Can Help

Population explosions in jellies are really a symptom of a much larger problem—today's oceans are suffering. In unhealthy oceans, fish decline while many jellies are prospering. You may not live anywhere near the ocean, but your choices can still make a huge difference in keeping our waters healthy for fish and jellies.

What can you do to help save our oceans? Take a look at the suggested books and websites listed at the back of this book.

*Swarms of sea nettles like these (*Chrysaora fuscescens*) from the Pacific coast of North America are a possible consequence of overfishing. When too many jelly predators are removed from the ocean, jelly populations can explode.*

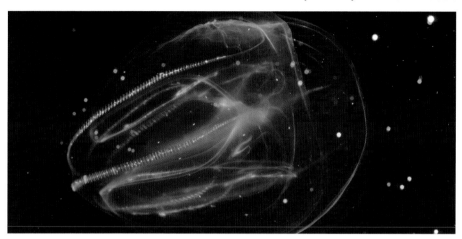

*This comb jelly (*Mnemiopsis leidyi*) can be found along much of the Atlantic coast of North and South America and has been introduced elsewhere. In Narragansett Bay, Rhode Island, warmer waters—a result of global warming and power plant discharge—have allowed jelly populations to multiply and extend their season. Jellies compete with juvenile fish for food: eggs, larvae, and copepods. The result? An overabundance of jellies and fewer fish.*

Phyllorhiza punctata *(above left) is an Indo-Pacific species that has been introduced into Hawaii and the western tropical Atlantic. It is an example of how introduced jellies can threaten the resources of an ecosystem. When comb jellies, like this* Beroe *sp. (above right), are accidentally introduced to a new environment, they compete with native animals for food.*

43

Glossary

Anemones (uh-NEH-muh-nees) — Jelly animals closely related to medusa jellies. Anemones are often brightly colored and have a polyp form that makes these animals look like underwater flowers. Most anemones live alone rather than in colonies.

Anthozoans (an-tho-ZO-uns) — A subgroup within Cnidaria made up of sea anemones, corals, and other jelly animals with a polyp form.

Asexual (AY-sex-you-ul) — Reproduction by division or splitting off from the parent. Having no reproductive organs.

Atrial Opening (AY-tree-ul) — Outgoing siphon for expelling water.

Bioluminescence (by-oh-LOO-muh-NES-ens) — A chemical process in living organ-isms that produces light.

Cilia (SILL-ee-ah) — Hairlike projections that sway in unison, creating movement through water

Cnidaria (ni-DA-ree-uh) — The scientific group to which medusa jellies, sea anem-ones, corals, and related animals belong. This group of animals was formerly called Coelenterata.

Colony — A group of animals, all of the same kind, that live together in a way that benefits each member of the group.

Crustacean (kra-STAY-shun) — Primarily aquatic and marine invertebrate animals that have segmented bodies, jointed legs, and a hardened external skeleton. Crabs, crayfish, shrimp, lobsters, barnacles, and water fleas are examples.

Ctenophora (teen-uh-FOUR-uh) — The sci-entific group to which comb jellies belong.

Cubozoans (kyu-bo-ZO-uns) — The name proposed by some scientists for a fourth Cnidarian group, which would include box-shaped jellies such as the sea wasp.

Gill — The respiratory organ in animals that obtains oxygen from the water and releases carbon dioxide. Most groups of marine animals have gills, including molluscs, crustaceans, and fish.

Gonad (GO-nad) — Reproductive organ. A group of male or female reproductive cells, which in jellies often line the sides of the stomach, but may extend through the bell of a jelly in the most mature specimens.

Hydra (HI-druh) — A kind of hydroid that lives only in fresh water. Hydra form colonies shaped like feathers or tiny trees.

Hydroid (HI-droid) — The polyp phase of a hydrozoan. Hydroids form colonies shaped like feathers or tiny trees.

Hydrozoans (hi-druh-ZO-uns) — A subgroup with-in Cnidaria made up of hydroids, hydra, and the Portuguese man-of-war. Hydrozoans can be either polyps or medusae.

Hypoxia (hi-POX-ee-ah) — Low oxygen saturation (levels) in the body.

Invertebrate (in-VERT-uh-brate) — An ani-mal without a backbone.

Lappet (LAP-it) — Lobes along the margin of the bell of a jelly.

Larva (LAR-vah): plural; larvae (LAR-vee) — An ani-mal's early life stage in which its appearance and way of life are very different from that of the adult.

Manubrium (man-OO-bree-um) — The tube between the stomach and the mouth of a jelly.

Medusa (muh-DOO-suh): plural; medusae (muh-DOO-see) — A bell- or umbrella-shaped form of Cnidaria. Adult moon jellies and sea wasps have a medusa form.

Nematocysts (NEM-at-uh -sists) — The sting-ing cells of jelly animals. Nematocysts are living weapons that are used both to capture food and for protection.

Ocellus (o-CELL-us) — The "eye spot" present in some jellies, capable of distinguishing light and dark.

Parasite (PAR-uh-site) — An animal that lives in or on other organisms, taking food from them but giv-ing nothing in return.

Photosynthesis (foh-toe-SIN-thuh-sis) — Process by which plants convert sunlight, car-bon dioxide, and water into energy in the form of sugars, starch-es, and other foods.

Phytoplankton (fi-tah-PLANK-ton) — Single-celled plant plankton. See zooplankton.

Plankton (PLANK-ton) — Algae and animals that drift with the surrounding water.

Poison — A toxin which is introduced to the body via the gastrointestinal tract or the respiratory tract.

Polyp (POL-up)—A form of Cnidaria made up of a soft stalklike body topped by a mouth surrounded by tentacles. Corals, sea anem-ones, and juvenile medusa jellies have a polyp form.

Rhopalium (row-PAH-lee-um): plural; rhopalia (row-PAH-lee-ah) — Specialized structures on some jellies (found along the margin of the bell) that con-tain one sensory organ for balance (statocyst) and another for distinguishing light and dark (ocellus).

SCUBA — Self-contained underwater breathing apparatus.

Scyphozoans (si-fuh-ZO-uns) — A subgroup within Cnidaria made up of moon jellies, sea wasps, and other jelly animals with a medusa form.

Siphonophore (si-FON-o-fore) -Hydrozoan jellies found in swimming or floating colonies composed of modified polyps and medusae.

Statocyst (STAT-oh-cist) — A balance organ, usually a calcium or magnesium carbonate crystal, the move-ment of which against surrounding cilia enables the medusa to determine its position in the water.

Strobilation (stroh-bill-AY-shun) — A form of asex-ual reproduction in some jellies in which miniature medusa-like structures are formed like stacked din-ner plates.

Symbiosis (sim-bee-OH-sis) — The living to-gether of two different organisms, in which one or both benefit from the relationship.

Tentacles (TEN-tuh-kuls) — Long flexible struc-tures around the mouths of medusae, sea anemones, and other jelly animals; used for grasping and stinging.

Tunicate (TOO-nik-it) —From the subphylum Tunicata and named for the tough cellulose sack that encases it, there are three classes of tunicates: Larvacea, Ascidiacea, and Thaliacea. Salps belong to the Thalicea.

Toxin (TOX-in) — A substance that is harmful to the tissues.

Venom — A toxin which usually enters the body by injection through intact skin (e.g.. a jelly sting).

Vertebrate (VER-tuh-brate) — An animal with a backbone.

Water column — The area of the ocean be-tween the surface and the bottom.

Zooplankton (zoh-uh-PLANK-tun) — Any floating animals (often microscopic) eaten by sea jellies

45

Further Reading

Books and References

Berril, Michael, and Deborah Berril. *The North Atlantic Coast: A Sierra Club Naturalist's Guide*. San Francisco: Sierra Club Books, 1981.

Broad, William J. *The Universe Below: Discovering Secrets of the Deep Sea*. New York: Simon & Schuster, 1998.

Burnett, Nancy, and Brad Matsen. *The Shape of Life*. Monterey, California: Monterey Bay Aquarium in association with Sea Studios, 2002.

Cairns, Stephen D. *Cnidaria and Ctenophora*. Bethesda, Maryland: American Fisheries Society, 1991.

Campbell, Eileen, illustrated by Andrea McCann. *A Guide to the World of the Jellyfish*. Monterey, CA: Monterey Bay Aquarium Foundation, 1992.

Connor, Judith, and Nora Deans. *Jellies: Living Art*. Monterey, CA: Monterey Bay Aquarium Foundation, 2002.

Halstead, Bruce W, M.D. *Poisonous and Venomous Marine Animals of the World*. Princeton, N.J.: The Darwin Press, 1988.

Halstead, Bruce W. M.D., Paul S. Auerback, M.D., and Dorman R. Campbell. *A Color Atlas of Dangerous Marine Animals*. Boca Raton: CRC Press, Inc., 1990.

Purcell, J.E., W.M. Graham and H. J. Dumont, editors. *Jellyfish Blooms: Ecological and Societal Importance*. New York, London: Kluwer Academic, 2000.

Ruppert, Edward, Robert D. Barnes, Richard S. Fox. *Invertebrate Zoology*, 7th ed. Stamford, CT: Thompson Learning, 2003.

Wrobel, David and Claudia Mills, *Pacific Coast Pelagic Invertebrates: A Guide to the Common Gelatinous Animals*. Monterey, CA: Monterey Bay Aquarium Foundation, 1998.

For Kids

George, Twig C. *Jellies: the Life of Jellyfish*. Brookfield, CT: Millbrook Press. 2000.

Glaspey, Marion. *Jellyfish and Kin of the Mid-Atlantic Coast*. South Orleans, MA: Arey's Pond Press, 1988.

Kovacs, Deborah, and Kate Madin. *Beneath Blue Waters: Meetings with Remarkable Deep-Sea Creatures*. New York: Viking, 1996.

Martin-James, Kathleen. *Floating Jellyfish*. Minneapolis: Lerner Publications Company, 2001.

McKenzie, Michelle. *Jellyfish: Inside Out*. Monterey, CA: Monterey Bay Aquarium Press, April, 2003.

Sharth, Sharon. *Sea Jellies: From Corals to Jellyfish*, NY, Toronto, London: Franklin Watts, a Division of Scholastic Inc., 2002.

Taylor, Leighton, and Norbert Wu. (eds.). *Jellyfish*. Minneapolis: Lerner Publications Company, 1998.

Videos

Jellies: Phantoms of the Deep. Aquarium of the Pacific, 2000.

Jellies and Other Ocean Drifters. A Sea Studios Production, Monterey Bay Aquarium. Narrated by Leonard Nimoy.

Websites for the world of jellies

British Marine Life Study Society
http://ourworld.compuserve.com/homepages/
BMLSS/Moonjell.htm

Hawaiian Lifeguards homepage
http://www.aloha.com/~lifeguards/jelyfish.html

The Jellies Zone http://jellieszone.com/index.html

Monterey Bay Aquarium http://www.mbayaq.org/

Monterey Bay Aquarium Research Institute
http://www.mbari.org/default.htm

National Aquarium in Baltimore
http://www.aqua.org/

New England Aquarium http://www.neaq.org

Pacific Coast Gelatinous Zooplankton
http://jellieszone.com/pacificjellies.htm

Scyphozoans Jellyfish
http://www.earlham.edu/~meckehe/
scyphozoajellyfish.htm

Zoo News Jellyfish Page
http://www.geocities.com/rabpid7/jellyfishx.html

Resources for Environmental Awareness

Books

Brower, M., and W. Leon. *The Consumer's Guide to Effective Environmental Choices: Practical Advice from the Union of Concerned Scientists.* New York: Three Rivers Press, 1999.

Dauncey. G., and P. Mazza. *Stormy Weather: 101 Solutions to Global Climate Change.* Gabriola Island: New Society Publishers, 2001.

Websites for Environmental Awareness

Through better shopping choices, you can help stop over fishing!

The world's ever-growing population loves seafood, and the oceans can't keep up with demand. Where and how a fish is caught can make a big difference in considering sustainability, as fisheries and fishing practices differ from place to place. Buying a variety of sustainably harvested fish minimizes pressure on a single fishery.

United States Environmental Protection Agency, Global Warming-Actions.

http://yosemite.epa.gov/oar/globalwarming.nsf/
content/Actions
IndividualMakeaDifference.html

Monterey Bay Seafood Watch Program
http://mbayaq.org/cr/seafoodwatch.asp>http://
mbayaq.org/cr/seafoodwatch.asp

Blue Ocean Institute
http://www.blueoceaninstitute.org/projects/
>http://www.blueoceaninstitute.org/projects/

Seafood Choices Alliance http://www.seafood-choices.com/>http://www.seafoodchoices.com/

The Ocean Project
http://www.theoceanproject.org/difference/>htt
p://www.theoceanproject.org/difference/

NEAq Fish of the Month Page
http://www.neaq.org/ecosound/index.php>http:/
/www.neaq.org/ecosound/index.php

Union of Concerned Scientists
http://www.ucsusa.org/global_environment/
global_warming/page.cfm?pageID=524

47

Photographic and Illustration Credits

All photos by David Wrobel except the following: p 6 Bottom left: Norman Katz; p 6 Bottom right: Kenneth Mallory, NEAQ; p 7 Top left: Paul Erickson, NEAQ; page 7 Bottom left: Paul Erickson, NEAQ; p 8 Top right: Art Resource Inc., NY, NY; p 9 Bottom left: Paul Erickson, NEAQ; p 9 Middle right: Ken Read; p 9 Bottom right: Norman Katz; p 10 Top right: Norman Katz; p 11 Bottom left: Norman Katz; p 13 Bottom right, Larry Madin; p 15 Top right: Norman Katz; p 15 Bottom right: Whitey Hagadorn, Amherst College; p 16: James Needham; p 17 Bottom left: Kenneth Mallory, NEAQ; p 17 right: Norman Katz; p 19 Top right: Kenneth Mallory, NEAQ; p. 20 Bottom left, Kenneth Mallory, NEAQ; p 21 Left: Norman Katz; p 22 Top left: J.F. Bertrand, NEAQ; p 22 Bottom: Kenneth Mallory, NEAQ; p 23: James Needham; p 26 Top right: Kenneth Mallory, NEAQ; p 27 Top: James Needham; p 29 Bottom left: Jamie Seymour, James Cook University; p 29 Bottom right: Norman Katz; p 30 Top right: Courtesy Center for Coastal Studies, Provincetown, MA; p 31 Top left: Kenneth Mallory, NEAQ; p 35 Bottom left: Kenneth Mallory, NEAQ; p 35 Bottom right: Larry Madin; p 38: Larry Madin; p 39 Bottom: Kenneth Mallory, NEAQ; p 41 Top left: Kenneth Mallory, NEAQ.

Acknowledgments

We would like to acknowledge the efforts of the many collaborators who made important contributions to this book. Principal author, Elizabeth Tayntor Gowell, has been unstinting in her work with the New England Aquarium, first as a former education staff member and Aquarium author, then as an independent author of several notable books for children. Aquarium staffer Cristina Santiestevan earns praise as the author of the book's final chapter, "Jellies and the Environment," reflecting the theme of the special New England Aquarium exhibit Amazing Jellies, which opened in April, 2004. She was assisted by Bonnie Epstein, the Principal Investigator for this exhibit; the entire book is indebted to Bonnie's review and suggestions. David Wrobel was the key photographer for this publication, which could not have been produced without his fine work. Photographer Norman Katz supplied many vital images. Science writer Melanie Perlman assisted with important captions. Aquarium designer Gretchen Mendoza helped with images. We are indebted to Woods Hole scientist Larry Madin for his critical eye in reviewing the text and for his help providing images. A special thanks go to the Monterey Bay Aquarium, especially Hank Armstrong and Nora Deans, for their help and support. We recommend their fine aquarium as a wonderful inspiration and resource about the world of water. Several scientists, in addition to Larry Madin, deserve recognition for contribution of images: Whitey Hagadorn from Amherst College; Jamie Seymour from James Cook University; and artist James Needham, who provided the wonderful illustrations. Finally, thanks to Ib Bellew, Carole Kitchel, Karen Palmer, and designer Louise Millar of Bunker Hill Publishing for making this happen on such short notice.

Ken Mallory, Editor-in-Chief
Publications Programs, New England Aquarium